SpringerBriefs in Molecular Science

Chemistry of Foods

Series editor

Salvatore Parisi, Industrial Consultant, Palermo, Italy

The series Springer Briefs in Molecular Science: Chemistry of Foods presents compact topical volumes in the area of food chemistry. The series has a clear focus on the chemistry and chemical aspects of foods, topics such as the physics or biology of foods are not part of its scope. The Briefs volumes in the series aim at presenting chemical background information or an introduction and clear-cut overview on the chemistry related to specific topics in this area. Typical topics thus include: - Compound classes in foods – their chemistry and properties with respect to the foods (e.g. sugars, proteins, fats, minerals, ...) - Contaminants and additives in foods – their chemistry and chemical transformations - Chemical analysis and monitoring of foods - Chemical transformations in foods, evolution and alterations of chemicals in foods, interactions between food and its packaging materials, chemical aspects of the food production processes - Chemistry and the food industry – from safety protocols to modern food production The treated subjects will particularly appeal to professionals and researchers concerned with food chemistry. Many volume topics address professionals and current problems in the food industry, but will also be interesting for readers generally concerned with the chemistry of foods. With the unique format and character of Springer Briefs (50 to 125 pages), the volumes are compact and easily digestible. Briefs allow authors to present their ideas and readers to absorb them with minimal time investment. Briefs will be published as part of Springer's eBook collection, with millions of users worldwide. In addition, Briefs will be available for individual print and electronic purchase. Briefs are characterized by fast, global electronic dissemination, standard publishing contracts, easy-to-use manuscript preparation and formatting guidelines, and expedited production schedules. Both solicited and unsolicited manuscripts focusing on food chemistry are considered for publication in this series.

More information about this series at http://www.springer.com/series/11853

Izabela Steinka · Caterina Barone
Salvatore Parisi · Marina Micali

The Chemistry of Frozen Vegetables

Springer

Izabela Steinka
Gdynia Maritime University
Gdynia
Poland

Caterina Barone
Associazione "Componiamo il Futuro"
  (COIF)
Palermo
Italy

Salvatore Parisi
Industrial Consultant
Palermo
Italy

Marina Micali
Industrial Consultant
Messina
Italy

ISSN 2191-5407   ISSN 2191-5415   (electronic)
SpringerBriefs in Molecular Science
ISSN 2199-689X   ISSN 2199-7209   (electronic)
Chemistry of Foods
ISBN 978-3-319-53930-0   ISBN 978-3-319-53932-4   (eBook)
DOI 10.1007/978-3-319-53932-4

Library of Congress Control Number: 2017932920

Printed on acid-free paper

This Springer imprint is published by Springer Nature
The registered company is Springer International Publishing AG
The registered company address is: Gewerbestrasse 11, 6330 Cham, Switzerland

Contents

Chapter 1
Antibiotic-Resistant Staphylococci Isolated from Hermetically Packaged Frozen Vegetables

Izabela Steinka, Caterina Barone, Salvatore Parisi and Marina Micali

Abstract The aim of this study was to evaluate the prevalence of antibiotic-resistant staphylococci in frozen vegetables during long-term storage of these products. The vegetables were subjected to freezing at −40 °C (fluidisation technique). Experimental temperature values in frozen tissues reached −18 °C. Vegetables were packed in polyethylene airtight bags. The study involved 33 assortments of frozen vegetables. The number of *Staphylococcus aureus* in frozen vegetables achieved 3.5 $\log_{10}$ CFU/g; 4.7% of these samples showed unacceptable microbial counts. The maximum level of methicillin-resistant *S. aureus* was 4.25 $\log_{10}$ CFU/g. All samples showed also the presence of aerobic bacteria without correlation between their number and staphylococci counts. On these bases, the Regulation (EC) No. 2073/2005, recently amended in 2007, could be modified with a condition concerning the absence of *S. aureus* in vegetable products ready for frozen storage because of the possible detection or the presence of methicillin-resistant *S. aureus* in final products.

Keywords Freezing · Frozen vegetables · Methicillin resistance · Staphylococci · *Staphylococcus aureus* · Yeasts and moulds

Abbreviations

CFU	Colony Forming Unit
MR	Methicillin resistant
MRSA	Methicillin-resistant *Staphylococcus aureus*
MSSA	Methicillin-sensitive *S. aureus*
MSSA	Methicillin-susceptible *S. aureus*
MDRA	Multidrug-resistant *S. aureus*
PE	Polyethylene
RH	Relative humidity

© The Author(s) 2017
I. Steinka et al., *The Chemistry of Frozen Vegetables*, Chemistry of Foods,
DOI 10.1007/978-3-319-53932-4_1

1.1 Introduction

Vegetable products are often correlated with the detection of varied microflora species (Hardy et al. 1977; Hassan et al. 2011; Johnston et al. 2005; Ofor et al. 2009; Tournas 2005a, b). The most frequently isolated microorganisms from vegetables are fungi belonging to the following types: *Alternaria, Fusarium, Cladosporium* and *Rhizopus* (Tournas 2005a). The autochthonous microflora of fresh vegetables are represented by *Erwinia* rod-shaped bacteria. Pathogenic species for human beings can also be found among bacteria. Fresh vegetables coming from trade can be identified as the source of life forms such as *Bacillus, Shigella, Clostridium* spp., *Escherichia coli* and *Staphylococcus aureus* (Ofor et al. 2009). However, the most commonly isolated bacteria from vegetables are rod-shaped *Pseudomonas* spp. (Barth et al. 2009). Reference books have shown that anaerobic *Clostridium* rod-shaped life forms can be also present among microflora isolated from tomatoes, mushrooms or cabbage (Barth et al. 2009; Buck et al. 2003). Soil and water can be regarded as extrinsic sources for the infection of vegetables with microorganisms. *E. coli* O157:H7 and *Salmonella* spp. have to be considered among the microorganisms whose source can be soil or water. The contamination of vegetables can occur as a result of using tools and appliances, in the process of collection and initial processing of plant raw materials. The presence and preservation of the *S. aureus* population in vegetables is omitted, although characteristics of contamination sources of vegetables with rod-shaped *Salmonella* and *Listeria* are mentioned in references books (Martin and Beelman 1996).

In addition, staphylococci are not frequently mentioned (Bernard et al. 1982; Fan and Sokorai 2007; Hardy et al. 1977; Manani et al. 2006; Silbernagel et al. 2003) during the assessment of the microbiological quality of frozen vegetables presented by certain authors.

When characterising the microflora of frozen vegetables, the authors of the studies draw attention mainly to microorganisms such as *Listeria monocytogenes* and *E. coli* (Fan and Sokorai 2007).

The early American studies regarding frozen vegetables showed that some processes, e.g. blanching, can contribute to the reduction of the level of microflora. However, although raw vegetable materials were blanched, the authors of these studies stated the presence of numerous microorganisms and also the presence of *E. coli* in 10% of tested products. In detail, the assessment of staphylococci counts in frozen vegetables showed the presence of *S. aureus* (Bernard et al. 1982). The most probable number of *S. aureus* isolated from frozen cauliflower ranged from 41 to 60 cells/g. In the case of frozen corn and peas, the most probable number of *S. aureus* did not exceed 20 cells/g (Bernard et al. 1982). The presence of *S. aureus* in vegetables was also recently reported (Viswanathan and Kaur 2001).

The presence of antibiotic-resistant microorganisms has been identified by several authors in raw materials and fresh food of animal origin (Can and Celik 2012; Adeshina et al. 2012; Hammad et al. 2012; Hassan et al. 2011; Nipa et al.

2011; Normanno et al. 2007; Rich and Roberts 2004; Steinka and Janczy 2013, 2014).

With relation to staphylococci coming from vegetables, there are few data regarding toxigenic properties and the resistance to antibiotics (Colombari et al. 2007; Adeshina et al. 2012; Igbeneghu and Abdu 2014; Nguyen-The and Carlin 2000; Nipa et al. 2011). As the source of staphylococci, the fresh products of plant origin are mentioned more frequently than the frozen ones. The presence of *Staphylococcus* spp. in carrots, lettuce and radish has been observed (Buck et al. 2003). Information on the presence of antibiotic-resistant bacteria in raw material of plant origin has also been reported recently (Gbonjubola 2011). Results of these studies showed that *S. aureus* appeared in up to 25% of tested samples of vegetable salads.

The assessment of antibiotic resistance was performed with the use of a diffusion method. The stated zones of inhibition of *S. aureus* by antibiotics were 9–25 mm for the following antibiotics: amoxicillin, augmentin, gentamicin, cotrimoxazole, nitrofurantonine and tetracyclines. Staphylococcal sensitivity was observed with ofloxacin, and the greatest resistance was reported in relation to nalidixic acid. The authors of the studies also stated the presence of 66.7% of *S. aureus* strains resistant to augmentin and over 39% of staphylococci resistant to gentamicin, tetracycline and cotrimoxazole (Adeshina et al. 2012).

Current data related to the identification of methicillin-resistant *Staphylococcus aureus* (MRSA) or multidrug-resistant *S. aureus* (MDRA) strains in frozen food of plant origin do not provide a complete knowledge on the safety of these products. There is also a lack of sufficient data on MRSA strains in frozen vegetables. Therefore, the aim of this study is the assessment of the occurrence of antibiotic-resistant staphylococci in frozen vegetables during the long-term storage of these products.

1.2 Materials and Methods

Frozen vegetables products have been treated with a fluidization method at −40 °C. Experimental temperature values in frozen tissues reached −18 °C. Vegetables were hermetically packaged in polyethylene (PE) pouches: 33 hermetically packaged frozen vegetables were examined. Over 40% of tested products consisted of one species of vegetable. These products included carrots, beans, spinach, Brussels sprouts, broccoli, beetroot, mushrooms, cauliflower and broad beans. Fourteen of these samples concerned multi-component mixed vegetables. These vegetables were kept in refrigerators at temperatures of −18 ± 2 °C (Table 1.1). Frozen vegetables came from producers representing 12 different brands present in 13 retail chains.

Table 1.1 Vegetable components and other ingredients in multi-component frozen products

Multi-component mixed vegetables	Vegetables
Mexican-style mixed vegetables	Carrot, French beans, corn, green peas, paprika, onion, celery, white mustard, soya bean, meal, milk
Italian -style mixed vegetables	Zucchini, French beans, carrot, cauliflower, broccoli
Chinese -style mixed vegetables	Carrot, bean Mung sprouts, paprika, leek, bamboo, 10% mushroom-style Mung
Mixed vegetables	Carrots, Brussels sprouts, green peas, cauliflower, French beans, parsley, celery, leek
Green *Romanesco* vegetables for a frying pan	30% *romanesco*-style cauliflower, broccoli
Caribbean-style vegetables for a frying pan	Broccoli, French bean, onion, carrot, pineapple, sauce (water, coconut milk, paprika, chilli, black pepper, curcuma, mango, cumin, coriander)
Assorted vegetables	Carrots, green peas, French beans, cauliflower
French-style vegetables for a frying pan	Cauliflower, carrot junior, green peas, sauce, butter, water, honey, potato starch, black pepper, parsley, nutmeg apple
Chinese vegetables for a frying pan	Carrots, leek, been Mung sprouts, red paprika, green peas, yellow carrots, sauce, water, rape oil, sugar, garlic, yellow carrots, soya bean, potato starch, ginger, black pepper, sanbal oil, lemon grass
Vegetables for frying pan with pasta and *romanesco* spices	French Bean, broccoli, corn, paprika, tomato, dry tomato, onion, 20% pasta, semmelin, water, 17% mushroom, 1% *Romanesco* spices, salt, garlic, sugar, sweet paprika, coriander, caraway, marjoram chilli, parsley, black pepper, eggs, milk, celery, white mustard soya bean
Vegetables for a frying pan	Fry potato, broccoli, carrot, French bean, corn, paprika, onion, eggs, milk, celery, white mustard, soya bean
Vegetables mix	Carrot, cauliflower, French bean, celery, parsley, kohlrabi, eggs, milk, white mustard, soya bean
Vegetables for a frying pan with dry tomatoes	Fry potato, French bean, paprika, tomato, dry tomato (3%), onion, mushrooms (15%), spices (1%)
Refinement-style vegetables for a frying pan	Carrot, French bean, paprika, parsley, onion, corn, leek, 35% white rice, 10% mushrooms, celery, eggs, grain
Red vegetables for a frying pan	Bean, paprika, tomato, carrot, onion, garlic
Romanesco-style red mix vegetables	Cauliflower, broccoli, 30% *romanesco*-style cauliflower
Flowers assorted vegetables	Carrot, cauliflower, broccoli
Soup vegetables	Carrot, parsley, leek, celery
Carrots with green peas	Carrot, green peas

The microbiological analysis included the determination of the total plate count of microorganisms, *Staphylococcus* spp. and *S. aureus* counts, and the count of fungi. Moreover, the presence of MRSA and related microbial count in 1 g of samples was determined.

1.2.1 Microbiological Analysis

Microbial counts for staphylococci were obtained in accordance with the norm PN-EN ISO 6888-2:2001 on Baird-Parker rabbit plasma fibrinogen agar (Merck). Total viable counts were obtained according to the methodology included in the norm PN-ENISO 4833-1:2013 with the pour-plate method (standard methods agar).

Fungi were determined on dichloran glycerol agar plates through 120-h incubation at 25 °C. Antibiotic resistance was assessed with the use of the Prolex™ StaphXtra Latex Kit (Biocorp. Polka Ltd.)

MRSA counts were obtained by means of the chromogenic medium for isolation and fast identification of methicillin-resistant *Staphylococcus* (Gross). Incubation was performed for 48 h at 37 °C. Pink-coloured colonies were qualified as MRSA, and isolated colonies were tested with the use of antibiotic-resistance tests.

Confirmation tests were performed in three stages:

- Stage 1. Assessment of microbiological quality through marking of the total cell count of mesophilic aerobic microorganisms, fungi and staphylococci in one- or multi-component products. This step concerned products of the four main producers present in the market
- Stage 2. Assessment of the occurrence of MRSA in products coming from eight brands manufactured for inexpensive retail chains
- Stage 3. Determination of the following counts for vegetable frozen products: MRSA, methicillin-susceptible *S. aureus* (MSSA) and *Staphylococcus* spp. methionine sulfoxide reductase.

1.3 Results

With reference to tested vegetable products, mesophilic aerobics at the level of 2.25–5.44 $\log_{10}$ CFU/g were found. The count of these bacteria in Brussels sprouts, spinach and carrots exceeded 4.0 $\log_{10}$ CFU/g on average. However, maximum values obtained for these vegetables exceeded this number and reached the value of 5.36 $\log_{10}$ CFU/g. Among the two- or four-component vegetables, 'carrots with peas' and 'soup vegetables' types showed the highest level for these microorganisms (Table 1.2).

The count of bacteria in 'carrots with peas' and mixed vegetables containing carrots, leek, celery and parsley root amounted, on average, to 5.32 and 3.85 $\log_{10}$ CFU/g, respectively (Table 1.2). A remarkable number of microorganisms was also

Table 1.2 Contamination ascribed to aerobic microorganisms and moulds in frozen vegetables from manufacturers I–IV (Fig. 1.2)

Number of range	Vegetable types	Counts of mesophilic aerobic microorganisms, average value	Mould counts
		Log_{10} CFU/g	Log_{10} CFU/g
1	French bean	2.55	0.0
2	Spinach	4.74	<1.0
3	Beetroot	2.75	0.0
4	Brussels sprouts	5.17	0.0
5	Carrots	4.65	0.0
6	Broccoli	3.87	<1.0
7	Broad bean	3.18	1.3
8	Cauliflower	4.28	0.0
9	Mushroom	3.87	0.0
10	Carrot with green peas	5.32	0.0
11	Assorted vegetables	3.85	0.0

Table 1.3 Contamination ascribed to aerobic microorganisms and moulds in frozen multi-component mixed vegetable products

Number of range	Vegetable types	Counts of mesophilic aerobic microorganisms, average value	Mould counts
		Log_{10} CFU/g	Log_{10} CFU/g
1	Mexican-style mixed vegetables	5.05	0.0
2	Italian-style mixed vegetables	3.36	1.48
3	Chinese -style mixed vegetables	5.39	1.48
4	Mixed vegetables	3.28	1.0
5	Green vegetables for a frying pan	5.04	0.0
6	Caribbean-style vegetables for a frying pan	4.02	1.3
7	French-style vegetables for a frying pan	2.60	0.0
8	Chinese-style vegetables for a frying pan	5.18	0.0
9	Red vegetables for a frying pan	3.97	0.0
10	Vegetables for a frying pan	3.69	0.0

stated in mixed vegetables selected for direct thermal processing. In Mexican-style and Chinese-style mixed vegetables, total cell counts were 5.05 and 5.39 $\log_{10}$ CFU/g on average respectively (Table 1.3).

The tests for the presence of filamentous fungi and yeast showed positive detection for fungi in seven products (Tables 1.2 and 1.3). Their counts did not exceed 1.48 $\log_{10}$ CFU/g. The highest value—3.85 $\log_{10}$ CFU/g of fungi—was stated in one type of vegetable only (soup vegetables). Fungi were absent in eight ranges of vegetables, and the average count of filamentous fungi in four product types did not exceed 10 CFU/g. A higher level of microbiological contamination was observed in the case of yeasts. A count exceeding 4.0 $\log_{10}$ CFU/g was stated in four situations (Fig. 1.1). Samples of spinach, cauliflower, mushrooms and Mexican-style mixed vegetables belong to this group. A high percentage of tested samples (52.4%) were characterised by the presence of yeast exceeding 2.0 $\log_{10}$ CFU/g (Tables 1.4 and 1.5), on average, of frozen vegetables.

S. aureus appeared in all tested one-component products. Related counts did not exceed 1.48 $\log_{10}$ CFU/g (Table 1.4). A higher level of *S. aureus* was stated in mixed vegetables. A high average count of staphylococci was also stated in the tested samples of mixed vegetables (Table 1.5).

Among the one-component products, the highest level of these microorganisms was observed in one of the spinach ranges (spinach leaves). A small count of cells of *S. aureus* was stated in the bouquet of floral vegetables from the first manufacturer. On the other side, frozen vegetables manufactured by the third manufacturer had the lowest percentage of samples showing the presence of *S. aureus* (Fig. 1.2).

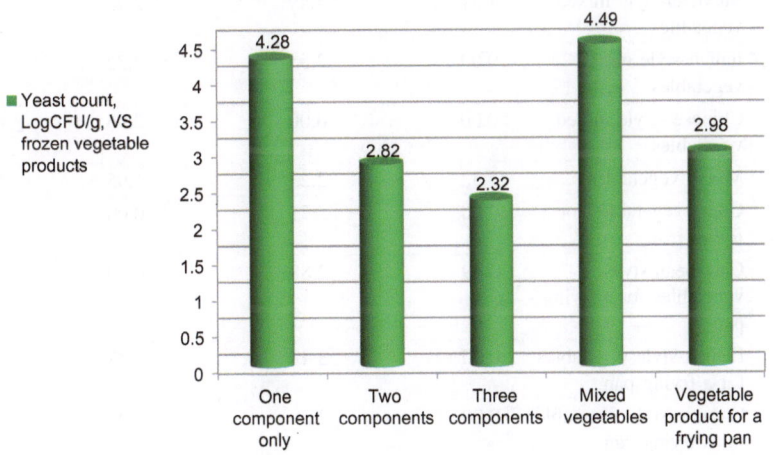

Fig. 1.1 Yeast contamination in frozen vegetable products. On the basis of available results, a relationship is established between yeast counts and the kind of vegetable products. In detail, yeast contamination depends on the quantity of vegetable components

Table 1.4 Microbial contamination ascribed to staphylococci and yeast in frozen products (manufacturers: I–IV) in one- and two-component products

Number of range	Vegetable types	S. aureus	Staphylococcus spp.	Yeasts
		Log_{10} CFU/g, average value	Log_{10} CFU/g, average value	Log_{10} CFU/g, average value
1	French bean	<1.00	2.66	1.12
2	Spinach	1.23	3.27	3.55
3	Beetroot	<1.00	2.27	1.42
4	Brussels sprouts	<1.00	3.31	2.20
5	Carrots	<1.00	3.25	2.10
6	Broccoli	1.30	2.66	1.48
7	Broad bean	<1.00	2.96	2.13
8	Cauliflower	<1.00	4.12	2.88
9	Mushroom	1.48	3.25	3.11
10	Carrot with green pea	1.30	4.20	1.41
11	Assorted vegetables	<1.00	2.15	0.00

Table 1.5 Microbial contamination ascribed to staphylococci and yeast in frozen products (manufacturers: I–IV) in multi-component mixed vegetables

Number of range	Vegetable types	S. aureus	Staphylococcus spp.	Yeasts
		Average value, Log_{10} CFU/g	Average value, Log_{10} CFU/g	Average value, Log_{10} CFU/g
1	Mexican-style mixed vegetables	1.31	4.49	2.79
2	Italian-style mixed vegetables	0.00	2.55	2.32
3	Chinese -style mixed vegetables	0.00	0.00	2.95
4	Mixed vegetables	3.52	2.23	1.95
5	Green vegetables for a frying pan	0.00	2.71	0.00
6	Caribbean-style vegetables for a frying pan	2.49	2.88	0.00
7	French-style vegetables for a frying pan	0.00	3.48	1.38
8	Chinese-style vegetables for a frying pan	0.00	2.2.	<1.00
9	Red vegetables for a frying pan	1.38	2.11	2.98
10	Vegetables for a frying pan	0.00	2.72	2.43

Staphylococci in Frozen vegetable Products (four manufacture types)

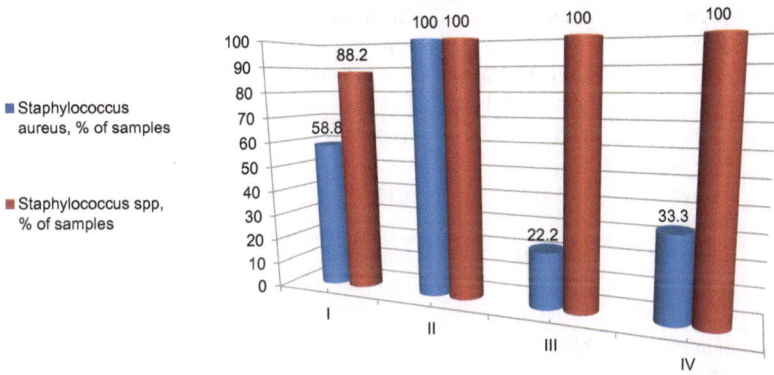

Fig. 1.2 Presence of *Staphylococcus aureus* and *Staphylococcus* spp. in frozen vegetables from different manufacturers

A high level of contamination with coagulase-negative staphylococci was stated in frozen vegetables. On average, their count ranged from 0.0 to 4.49 $\log_{10}$ CFU/g (Tables 1.4 and 1.5).

The minimum and the maximum observed values for the population of *Staphylococcus* spp. were compared in all tested ranges of frozen vegetables from manufacturers I–IV (Table 1.6). The composition of mixed vegetables and the conditions of the technological process involved a probable factor determining the presence of these bacteria.

Table 1.6 Maximum and minimum contamination ascribed to staphylococci in frozen vegetables in function of the manufacturer

Microbial contamination	Plant	Minimum count	Maximum count
		$\log_{10}$ CFU/g	$\log_{10}$ CFU/g
S. aureus	I	1.00	2.00
Staphylococcus spp.		1.00	4.36
S. aureus + *Staphylococcus* spp.			>4.00
S. aureus	II	1.00	2.04
Staphylococcus spp.		2.66	4.49
S. aureus + *Staphylococcus* spp.			>4.00
S. aureus	III	1.48	3.50
Staphylococcus spp.		1.95	4.26
S. aureus + *Staphylococcus* spp.			>4.00
S. aureus	IV	1.30	2.49
Staphylococcus spp.		2.11	3.27
S. aureus + *Staphylococcus* spp.			>4.00

Table 1.7 Microbial counts ascribed to staphylococci and MRSA detection in selected products of brands available in inexpensive retail chains

Number of range	Vegetable types	S. aureus	Staphylococcus spp.	MRSA	Manufacturer
		Log_{10} CFU/g range of values	Log_{10} CFU/g Range of values		
1	Green pea with carrot	0.00	0.00	−	V
2	Carrots junior	2.43–3.17	3.17	−	VI
3	Spinach	1.77–2.09	1.00–3.18	+	V
4	Spinach '1000 leaf'	2.00	1.23	+	VII
5	Spinach 'brick'	2.78–3.07	1.23	+	VIII
6	Brussels sprouts	0.00	0.00	−	V
7	Broccoli	2.50–2.77	2.59	+	IX
8	Green Pea	0.00	1.00	−	X
9	French bean	0.00	1.77	−	XI
10	Corn	2.02–2.97	1.86	+	X
11	Cauliflower	2.13–2.47	1.47	+	XII
12	French bean	0.00	0.00	−	X

With relation to MRSA, '+' is for 'presence'; '−' is for 'absence'

Other tests were carried out for the assessment of the survival of staphylococci and MRSA in mixed vegetables used samples from eight brands (trade numbers: V–VIII) present in the market and available in inexpensive retail chains (Table 1.7). Methicillin-resistant (MR) staphylococci were stated in 50% of manufacturers producing vegetables for inexpensive retail chains (trade numbers: V–VIII). Broccoli and spinach showed MRSA presence, similarly to other products containing spinach. Moreover, products of six out of eight brands available in inexpensive retail chains (tested for the presence of antibiotic-resistant staphylococci) showed the presence of MRSA strains (Table 1.7).

With relation to the verified MRSA detection in the majority of vegetable products, the assessment of antibiotic-resistant staphylococci counts in selected products was performed (Tables 1.8 and 1.9). Related results showed that MRSA were present in the majority of one-component product ranges (Table 1.8). MRSA counts ranged from 0.0 to 4.25 log_{10} CFU/g. The average microbial count in vegetables was 3.06 log_{10} CFU/g. In addition, MRSA were not stated in five products: yellow beans, Brussels sprouts, red beetroot, broccoli and spinach leaves (Table 1.8). The remaining ranges of spinach were characterised by the presence of both coagulase-positive and coagulase-negative staphylococci. The one-component

Table 1.8 Microbial contamination ascribed to MR staphylococci in one-component vegetable frozen products

Number of range	Vegetable types and manufacturer identification, where possible	Methicillin-resistant *S aureus* (MRSA)	MR *Staphylococcus* spp.
		Average value, Log_{10} CFU/g	Average value, Log_{10} CFU/g
1	Yellow French bean I	0.00	0.00
2	Green French bean I	3.04	0.00
3	Yellow Bean II	0.00	1.47
4	Green French bean II	2.14	0.00
5	Green French bean III	0.00	0.00
6	Broccoli I	3.56	1.60
7	Broccoli II	0.00	0.00
8	Brussels	0.00	0.00
9	Spinach leaf I	0.00	0.00
10	Spinach leaf II	0.00	0.00
11	Crumble spinach	4.25	3.85
12	Spinach 'chaplets'	1.77	1.00
13	Spinach leaf III	3.61	0.00
14	Cauliflower	2.00	2.68
15	Beetroot	0.00	1.60

frozen products showed the presence of coagulase-negative *Staphylococcus* spp. The average count of these bacteria in vegetables reached 2.71 log_{10} CFU/g.

Also, the presence of antibiotic-resistant coagulase-positive and coagulase-negative staphylococci was stated in multi-component mixed vegetables (Table 1.9). The average count of *S. aureus* and *Staphylococcus* spp. in these products did not exceed 1.56–1.54 log_{10} CFU/g; this number was lower than stated values for one-component frozen vegetables. The presence of *Micrococcus* spp. was stated in over 41% of frozen mixed vegetables. The count of MR *Micrococcus* spp. did not exceed 2.77 log_{10} CFU/g (Table 1.9).

The count of the aerobic microflora population mentioned by some authors for fresh vegetables differs from obtained values for frozen vegetables. This result depends on vegetable collection procedures, preservation technology and storage conditions in the freezing state.

The linear dependence between storage times in a cooling condition and microbial levels found on the surface of fresh vegetables was stated (Zhuang et al. 2003). Aerobic mesophilic bacterial counts in fresh vegetables were assessed by these authors to be 4–6 log_{10} CFU/g. The presence of these microorganisms in frozen vegetables indicates the participation of microflora resistant to low temperatures during the freezing process.

The level of aerobic mesophilic microflora present in frozen vegetables obtained in minor studies is compliant with results obtained by other researchers (Heard

Table 1.9 MRSA counts and microbial contamination ascribed to MR *Staphylococci* and MR *Micrococcus* spp. in multi-component vegetable frozen products

Multi-component products	MRSA	MR *Staphylococcus* spp.	MR *Micrococcus* spp.
	Log_{10} CFU/g	Log_{10} CFU/g	Log_{10} CFU/g
Spring assorted vegetables	0.00	0.00	0.00
Flower assorted vegetables	0.00	0.00	0.00
Chinese-style vegetables	1.60	1.84	0.00
Flower assorted vegetables	0.00	0.00	0.00
Soup vegetables	0.00	0.00	0.00
French-style vegetables for a frying pan	1.00	1.90	2.77
Chinese-stylevegetables for a frying pan	1.84	2.07	2.14
Vegetables for a frying pan with pasta and *romanesco* spices	1.30	1.30	2.79
Vegetables for a frying pan	2.20	2.00	1.60
Vegetable mix	0.00	0.00	0.00
Assorted Vegetables	0.00	0.00	0.00
Vegetables for a frying pan with dry tomatoes	1.47	1.30	0.00
Mexican-style mix	0.00	0.00	1.47
Italian-style mix	1.84	0.00	0.00
Romanesco-style green mix	1.00	0.00	0.00
Romanesco-style red mix	0.00	0.00	0.00
Refinement-style vegetables for a frying pan	1.00	1.00	1.30
Vegetable mix	1.30	0.00	2.25

2000). In particular, the presence of aerobic microorganisms within the range of 4–9 log_{10} CFU/g was reported. Our previous studies showed that the level of these microorganisms was 4.0 log_{10} CFU/g on average, (Steinka 2012).

The dynamics of the development of microorganisms in hermetically packaged frozen vegetables depends on many factors: among others, air amount and packaging barrier properties have to be considered. With concern to aerobic microorganisms, the most frequent types are rod-shaped bacteria belonging to *Pseudomonas* spp. because of their ability to survive at low temperatures and in presence of pectinolytic enzymes.

Amounts of fungi found by some authors in fresh vegetables are specified within the range of 7.3–16.0 log_{10} CFU/g (Tournas 2005a). Among the microflora having an impact on the quality of vegetables and fruits, the most important role is played by filamentous fungi *Penicillium, Aspergillus, Botritis, Rhizopus, Candida, Rhodotorula* and *Kloeckera* spp. (Nguyen-The and Carlin 2000).

Mould counts in one-component frozen products such as carrots and leek exceeded 2.0 $\log_{10}$ CFU/g in our previous studies (Steinka 2012). In detail, the level of mould did not exceed 2.25 $\log_{10}$ CFU/g: this maximum value may indicate the high quality of raw materials used for production of these frozen products or the effectiveness of freezing technology for stopping the development of these microorganisms.

On the other side, the amount of yeasts obtained in our earlier studies for one-component products ranged from 1.86 to 3.41 $\log_{10}$ CFU/g. It should be highlighted that this level concerned products immediately after freezing and not the ones that were stored (Steinka 2012). Actions such as peeling, cutting, washing and dewatering have an impact on the spoilage of vegetables. During the process of preparing vegetables for freezing, bacterial contamination can transferred from hands and gloves (Barth et al. 2009). Employers' hands and gloves used during the initial processing of vegetables can be a significant source of staphylococci.

Studies of some authors showed that the presence of *S. aureus* was stated in 11.8% of samples concerning fresh vegetables (Hammad et al. 2012). Detailed studies conducted by Colombari and co-workers showed that strains isolated from vegetable salads coming from five food handlers characterised both the same phage type and the profile of antibiotic resistance (Colombari et al. 2007).

These studies confirmed the possibility of *S. aureus* survival in vegetables after the freezing process. Similar data regarding the presence of *S. aureus* in fresh vegetables were submitted in the report of studies conducted in Australia (Department of Health 2006). The presented report showed that the presence of *S. aureus* was stated in all samples of fresh vegetables. However, quantitative results obtained for these studies showed that the level of staphylococci did not exceed 10^2 CFU/g. This level is also comparable with our data and presented in this work. Tested frozen vegetables showed counts ascribed to staphylococci can reach 3.5 $\log_{10}$ CFU/g. However, this result is not compliant with data described by us in an earlier publication (Steinka 2012). These studies showed that staphylococci were not present in the products after the freezing process of vegetables. This difference can indicate a crucial role of sub-lethal damages suffered by staphylococci cells; consequently, suitable conditions for proper product storages can be chosen on these bases.

The results of some studies showed also that antibiotic-resistant strains of *Staphylococcus* spp. could be stated in over 10% of the tested vegetables available on the market (Nipa et al. 2011).

In addition, recent studies of fresh vegetables conducted by some authors with concern to the possible detection of staphylococcal enterotoxins showed that 73.1% of *S. aureus* samples had the ability to synthesise *Staphylococcus* enterotoxin type A. A small percentage of isolated bacteria were able to synthesise enterotoxins A + B and A + C (Moon et al. 2007). This behaviour can be conditioned by the ability of *S. aureus* to grow and synthesise enterotoxin A in hermetically packages with low oxygen content (Martin and Beelman 1996).

Reference books' data show that blanched vegetables soon after the process of freezing should not have a population bigger than 1000 cells/g (Jay et al. 2005).

Initial processing steps for these products, such as cutting and slicing, can be a potential source of re-infection of raw vegetables. This hypothesis was confirmed in recent researches (Garg et al. 1990; O'Brien et al. 2000). Reported data showed stages of the technological process that can increase by seven times the level of microflora in plant tissues. On the basis of these data, it can be assumed that blanching should contribute to a reduction of the level of microflora by 3–3.6 $\log_{10}$ CFU/g (Splittsoesser et al. 1980). According to these authors, microflora of packaged vegetables should not exceed the level of 4.72–5.94 $\log_{10}$ CFU/g.

Some authors reported also that the re-packing process of vegetables can contribute to re-infection (Manani et al. 2006; Tournas 2005a). Among the isolated bacteria, *Pseudomonas*, *Escherichia* and *Bacillus* spp. were found; *S. aureus* was also identified during re-packing. Authors of these studies confirmed the presence of enterotoxigenic staphylococcal strains in re-packed products. In all of the samples, the presence of enterotoxin A, and in some samples also the presence of enterotoxins B and C, was stated. However, presented data concerned only fresh vegetables.

It is known that the superficial charge of various parts of vegetable raw materials is significant for forming pathogenic biofilms (Buck et al. 2003). However, few data regarding the preservation of pathogenic microflora in the biofilm formed on the surface of fresh vegetables are available. This knowledge is significant for assessing to what extent phytopathogenic microflora contributes to the growth of pathogens, including staphylococci.

In our studies, the presence of *S. aureus*, and also of *Staphylococcus* spp., was stated in frozen products. The count of these bacteria was up to 3.85 $\log_{10}$ CFU/g. Their presence was probably the result of hygienic negligence during collecting or freezing processes of vegetables. However, contamination factors of vegetables are limited to a small group everywhere: soil particles, air borne spores and water irrigation. Staphylococcal infection can occur while a vegetable is growing in the field as well as during postharvest storage and processing (Barth et al. 2009; FAO/WHO 2008).

Reported data show that *Staphylococcus* spp. strains isolated from plant products can also manifest antibiotic resistance. Three *Staphylococcus* spp. strains were isolated from fresh vegetables resistant to popular antibiotics except for chloramphenicol (Nipa et al. 2011).

Blanching—which precedes the freezing process of vegetables—should support a reduction of microflora counts in vegetables. Freezing should reduce the microbial count remaining after initial processing and blanching. It is probable that the low effectiveness of blanching or poor hygienic conditions were the reasons for the presence of these staphylococci in the products tested by us.

Vegetables are most often colonised by bacteria after tissues are damaged, because of the presence of a cuticle layer in undamaged materials. Conditions that lead to colonisation of vegetable surfaces by bacteria have been recently discussed (Mandrell et al. 2006). Some of these conditions are as follows: bacterial recognition of surfaces, and conditions for superficial adhesion.

Our results suggest that a significant factor of bacterial colonisation of vegetables can also be the occurrence of microscopic leakage in vegetable tissues. Hence, the possibility of the superficial development of staphylococci in frozen vegetables can determine the effectiveness of blanching as well as of storing in a frozen state. Minor tissue damages resulting from the creation of ice crystals during freezing can contribute to staphylococcal colonisation of vegetable tissue surfaces.

Another problem related to the presence of microorganisms in frozen vegetables results from the fact that bacterial biofilms formed on vegetable tissues with some microscopic damage are more difficult to remove than those formed on undamaged surfaces.

The freezing process can destroy external cell layers of the biofilm formed on the tissue surface. Staphylococci isolated after freezing and stored in this state come from the external layers of the biofilm. During storage, these cells are covered by an external biofilm layer of bacteria died during freezing. Staphylococcal cells attached to the exopolysaccharide layer from damaged tissues have an opportunity to survive for a long-term period of storage in the frozen state.

Studies of some authors showed that the biofilm formed by *S. aureus* is stimulated by the glycan of β-1,6-linked 2-acetamido-2-deoxy-D-glucopyranosyl residues with a net positive charge that promotes intercellular aggregation and attachment of cells to the vegetable tissue surface (Rohde et al. 2010). The lack of reduction of *S. aureus* counts on the surface of frozen vegetables was observed in the model conditions of experiments (Silbernagel et al. 2003). However, biofilm formation by means of the production of polysaccharide intercellular adhesin was not found for MRSA (O'Neill et al. 2007).

In the case of these studies, one should conclude that exopolysaccharides coming from micro-damaged tissues are necessary for MRSA to survive on vegetable surfaces. This type of interaction between the tissue of frozen vegetables and staphylococci is the most probable cause of MRSA infection in our studies.

There are data on the isolation of MRSA from fresh products of plant origin in reference books. Recent researches showed that MRSA can be present in fresh lettuce (Jung et al. 2000). The presence of *S. aureus* was stated in up to 30% of pumpkin samples, and 80% of these products revealed the presence of *Staphylococcus* spp. strains (Igbeneghu and Abdu 2014). *S. aureus* isolated from this vegetable was resistant to many antibiotics: among others, cephalosporin, chloramphenicol, gentamicin, tetracycline, ciprofloxacin, ofloxacin have to be mentioned in this ambit. Cephalosporin and trimethoprim were ineffective against 100% of *S. aureus* strains isolated from pumpkin. The authors of these studies thought that one of the reasons for the presence of *S. aureus* and *S. epidermidis* in these products was the lack of hand hygiene of workers in the room. 50% of tested *S. epidermidis* for sensitivity to antibiotics showed resistance to cephalosporin, chloramphenicol and trimethoprim among other antibiotic agents.

The state of staphylococcal contamination of selected vegetables was assessed (Moon et al. 2007). Their studies showed that *S. aureus* was present in 15% of sprouts samples, 13% of spinach and 4% of Chinese cabbage. Researchers stated that 73.1% of vegetables showed the presence of staphylococci. These strains can produce coagulates and type E-staphylococcal enterotoxins. The authors stated that the sensitivity of isolated strains depended on the source of microbial isolation.

Obtained results showed that bacteria isolated from vegetables had a higher resistance to penicillin and ampicillin. Up to 94% of the tested strains did not show sensitivity to these two antibiotics, while 10% were resistant to gentamicin and 8% to erythromycin. The reason for antibiotic resistance of bacteria isolated from vegetables is seen in the improper farming policy, in the non-compliance with a period of grace after the treatment of breeding animals and in the presence of pathogens in people employed in breeding farms and plant cultivation (Khachatourians 1998).

Reference books show that 81.25% of isolated fresh vegetables such as tomatoes, cucumbers, carrots and beetroot were characterised by the presence of antibiotic-resistant *Staphylococcus* spp. strains. It was stated that 11 of the tested strains were resistant to all antibiotics (Nipa et al. 2011).

Moon and co-workers stated that fresh vegetables are the main source of *S. aureus* (Moon et al. 2007) resistant to penicillin and ampicillin. Up to 94% of sampled vegetables showed the presence of microorganisms resistant to these antibiotics. 15% of staphylococci isolated from these vegetables were resistant to tetracycline.

Staphylococcal resistance to low temperatures can be another reason for their presence in vegetable products after the freezing process. There are few reports regarding the presence of staphylococci resistant to antibiotics in frozen vegetables. Studies on the preservation of staphylococci in frozen vegetables in the model system were recently published (Silbernagel et al. 2003).

During the assessment of survival of *S. aureus* in frozen mixed vegetables, it was stated that staphylococci can survive the preservation process. The referent strain was *S. aureus* ATCC 25923 with an inoculum from 2.0 to 4.0 $\log_{10}$ CFU/g. Vegetable samples were frozen to -15 °C and then stored for 7 days (Silbernagel et al. 2003). Obtained results confirmed that the level of staphylococcal cells surviving the freezing process depends on the initial count but also on the level background. Studies also showed the presence of *S. aureus* between 2.4 and 2.99 $\log_{10}$ CFU/g, for an inoculum of 2.0–3.0 $\log_{10}$ CFU/g, in 92.3% of samples after the process of storage. Only 7.7% of the tested samples showed less than 1.0 $\log_{10}$ CFU/g of staphylococci.

A similar low level of reduction in the count of staphylococci was achieved with inoculums of 3.0–4.0 $\log_{10}$ CFU/g in the model without a microflora background. In this variant, the count ranged from 3.0 to 4.8 $\log_{10}$ CFU/g in all samples. In the case of the presence of a microflora background of the same level, no significant reduction in the count of *S. aureus* in frozen vegetables was stated. The count of staphylococci reached the minimum of 3.11 $\log_{10}$ CFU/g and the maximum of 3.96 $\log_{10}$ CFU/g in 96.2% of samples.

Results of the quoted studies confirm our thesis that the process of freezing does not significantly reduce the count of staphylococci in vegetables, and that regardless of whether there are also other microorganisms present, their count does not significantly affect the preservation of staphylococci in these products.

Our studies showed the presence of MRSA, MSSA and *Micrococcus* spp. The count of these latter microorganisms reached the level of 2.79 $\log_{10}$ CFU/g. This fact can also be an issue of concern about the safety of consumers of vegetable products that are not subject to thermal processing before consumption. Apart from the data presented in model conditions by Sibelnagel and co-workers, there is a lack of information on antibiotic-resistant bacteria in frozen vegetables (Silbernagel et al. 2003).

The mechanisms of acquiring resistance to antibiotics most frequently include a horizontal gene transfer. Through conjugation, transformation and transduction, the *mecA* gene responsible for resistance of staphylococci can be transferred to other cells. Resistance to methicillin is associated with acquisition of a mobile genetic element from staphylococci. This element contains resistance genes such as *mecA* that encode new penicillin-binding proteins. A protein named PBP2a is resistant to inhibition by β-lactams (Džidic et al. 2008).

Soil contamination with animal faeces treated or accompanied with antibiotics can be one of the reasons for the occurrence of these bacteria in vegetables. Circulation of these bacteria in dust particles and water constitute a reason for their later presence on the surface of plant tissues. Improperly conducted sanitation of vegetables collected from such an environment contributes to the fact that these microorganisms remain on the surface. Damage of the tissue of vegetables that result from formation of small ice crystals during the process of freezing can contribute to the fact that microorganisms can more easily access the substances under the superficial layer of vegetable tissues. Some part of the MRSA population dies during freezing and these dead cells are the source of genes responsible for antibiotic resistance. In this situation, transformation is a mechanism stimulating the acquisition of antibiotic resistance.

Data presented by some authors showed that different species of staphylococci isolated from food can be found. At least 11 strains resistant to all of the most important antibiotics can be found: this delivery confirms the possibility of a transfer of the resistance gene among staphylococci.

The transfer of antibiotic resistance among bacteria as a phenomenon during cultivation was monitored (Khachatourians 1998). However, methicillin resistance is not the only type of reaction of staphylococci to the presence of antibiotics (Giedraitiene et al. 2011). The mechanisms in *S. aureus* cells allowing for the enzymatic inactivation of such antibiotics as aminoglycosides were stated. A lot of data presented in reference books show that a common bacterial mechanism of acquiring resistance to antibiotics is to use pumps that do not allow the antibiotic to enter the cell (Alekshun and Levy 2007). This mechanism is used by *S. aureus*, among others, for resistance to tetracycline, ciprofloxacin and norfloxacin.

The occurrence of antibiotic-resistant staphylococci in food of plant origin does not raise an issue of concern unless their count is outside specified limits. The count of coagulase-positive *S. aureus* in ready-to-eat food classifies this type of food in four categories, according to Australian standards (Department of Health 2006). Contamination is defined as a marginal count of staphylococci not exceeding 10^2–10^3 CFU/g. The third category applies when the count of coagulase-positive staphylococci is present at the unsatisfactory level of 10^3–10^4 CFU/g. The assessment of food specified as potentially hazardous is identified by a count higher or equal to the count of coagulase-positive and enterotoxin-positive *S. aureus*.

According to these standards, the majority of vegetables tested by us should be included in the products containing a so-called marginal level of staphylococci. Our studies showed that during the process of storing frozen of vegetables, no significant reduction of staphylococci was observed. The presence of antibiotic-resistant *Staphylococcus* spp. strains that cannot synthesise coagulates and the presence of a small count of antibiotic-resistant *Micrococcus* spp. was stated.

An attempt to assess the safety of food where the presence of antibiotic-resistant bacteria is stated was made as a specific index, but it did not regard fresh vegetables (Adeshina et al. 2012). Perhaps, further studies should be conducted for identifying such an index for antibiotic-resistant microflora that occurs in frozen vegetables.

1.4 Conclusions

The count of *S. aureus* in frozen vegetables ranges from 0.0 to 3.5 $\log_{10}$ CFU/g. 4.7% of examined samples had unacceptable levels of these microorganisms. The products manifested the presence of MRSA in 48.4% of the proposed ranges. The maximum MRSA count was 4.25 $\log_{10}$ CFU/g and maximum level antibiotic-resistant *Staphylococcus* spp. was 3.85 $\log_{10}$ CFU/g. Out of 100% of the samples, aerobic bacteria were also isolated. However, no correlation between their presence and fungi counts on the one side and the level of staphylococci on the other hand was stated. Hence, the possibility of the development of staphylococci on the surface of frozen vegetables can determine the effectiveness of the blanching process as well as of storing in a frozen state. Minor damage of the tissue resulting from the creation of ice crystals during freezing can contribute to staphylococcal colonisation of vegetable tissue surfaces. The Regulation (EC) No. 2073/2005, amended in 2007, should include the following condition: that *S. aureus* should not appear in vegetable products that are to be frozen, because the presence of *S. aureus* can suggest possible MRSA detection.

References

Adeshina GO, Jibo SD, Agu VE (2012) Antibacterial susceptibility pattern of pathogenic bacteria isolates from vegetable salad sold in restaurant in Zaria, Nigeria. J Microbiol Res 2(2):5–11. doi:10.5923/j.microbiology.20120202.02
Alekshun MN, Levy SB (2007) Molecular mechanisms of antibacterial multidrug resistance. Cell 128(6):1037–1050. doi:10.1016/j.cell.2007.03.004

Barth M, Hankinson TR, Zhuan H, Breidt F (2009) Microbiological spoilage of fruits and vegetables. In: SperberWH, Doyle MP (eds) Compendium of the microbiological spoilage of foods and beverages, food microbiology and food safety. Springer, New York.doi:10.1007/978-1-4419-0826-1

Bernard RJ, Duran AP, Swartzentruber A, Schwab AH, Wentz BA, Read RB (1982) Microbiological quality of frozen cauliflower, corn, and peas obtained at retail markets. Appl Environ Microbiol 44(1):54–58

Buck JW, Walcott RR, Beuchat LR (2003) Recent trends in microbiological safety of fruits and vegetables. Plant Health Progress 10:1092–1098. doi:10.1094/PHP-2003-0121-01-RV

Can HY, Celik TH (2012) Detection of enterotoxigenic and antimicrobial resistant *S. aureus* in Turkish cheeses. Food Control 24(1–2):100–103. doi:10.1016/j.foodcont.2011.09.009

Colombari V, Mayer V, Laicini DB, Zaira M, Mamizuka E, Franco BDGM, Destro T, Landgraf M (2007) Foodborne outbreak caused by *Staphylococcus aureus*: phenotypic and genotypic characterization strains of food and human sources. J Food Prot 70(2):489–493

Department of Health (2006) Microbiological quality of fruit and vegetables in Western Australian retail outlets 2005. Results of pilot survey designed by the Western Australian Food Monitoring Program (WAFMP) to establish data on microbiological quality of range of raw fruit and vegetables, domestic and imported. Government of Western Australia, Department of Health, Perth. Available https://www.health.wa.gov.au/publications/documents/WAFMP%20Technical%20report_Microbiological%20quality%20of%20Fruit%20&%20Veg_Final%20version%2060511.pdf. Accessed 18 Oct 2016

Džidić S, Šušković J, Kos B (2008) Antibiotic resistance mechanisms in bacteria: biochemical and genetic aspects. Food Technol Biotechnol 46(1):11–21

Fan X, Sokorai KJ (2007) Effect of ionizing radiation on sensorial, chemical and microbiological quality of frozen corn and peas. J Food Prot 70(8):1901–1908

FAO/WHO (2008) Microbial hazards in fresh fruit and vegetables. Microbiological Risk Assessment Series, Food and Agriculture Organization of United Nations (FAO) and the World Health Organization (WHO), pp 1–28

Garg N, Churey JJ, Splistoesser DF (1990) Effect of processing conditions on the microflora of fresh–cut vegetables. J Food Prot 53(8):701–703

Giedraitiene A, Vitkauskine A, Naginiene R, Pavilonis A (2011) Antibiotic resistance mechanisms of clinically important bacteria. Medicina (Kaunas) 47(3):137–146

Hassan SA, Altalhi AD, Gherbawy YA, El-Deeb BA (2011) Bacterial load of fresh vegetables and their resistance to the currently used antibiotics in Saudi Arabia. Foodborne Pathog Dis 8(9):1011–1018. doi:10.1089/fpd.2010.0805

Hammad AM, Watanabe W, Fuji T, Shimamoto T (2012) Occurrence and characteristics of methicilin–resistant and susceptible *Staphylococcus aureus* and methicilin-resistant coagulase negative staphylococci from Japanese retil ready-to eat raw fish. Int J Food Microbiol 156(3):286–289. doi:10.1016/j.ijfoodmicro.2012.03.022

Hardy D, KraegerSJ Dufour SW, Cady P (1977) Rapid detection of microbial contamination in frozen vegetables by automate impendence measurements. Appl Environ Microbiol 34(1):14–21

Heard G (2000) Microbial safety of ready-to-eat salads and minimally processed vegetables and fruits. Food Sci Technol Today 14:15–21

Igbeneghu OA, Abdu BA (2014) Multiple antibiotic–resistant bacteria on fluted pumpkin leaves, a herb of therapeutic value. J Health Popul Nutr 32(2):176–182

Jay MJ, Loessner MJ, Golden DA (2005) Modern food microbiology. Springer, New York

Johnston LM, Jaykus LA, Moll D, Martinez MC, Anciso J, Mora B, Moe C (2005) A field study of the microbiological quality of fresh produce. J Food Prot 68(9):1840–1847

Jung HJ, Cho JIL, Park SH, Ha SD, Lee KH (2000) Genotypic and phenotypic characteristic of *Staphylococcus aureus* from lettuces and raw milk. Korean J Food SciTechnol 37:134–141

Khachatourians GG (1998) Agricultural use antibiotics and the evolution of transfer of antibiotic-resistant bacteria. Can Med Assoc 159(9):1037–1129

Manani TA, Collison EK, Mpuchane S (2006) Microflora of minimally procesed frozen vegetables sold in Gaborne. Bostwana. J Food Prot 69(11):2581–2586

Mandrell RE, Gorski L, Brandl MT (2006) Attachment of microorganisms to fresh produce. In: Sapers GM, Gorney JR, Yousef AE (eds) Microbiology of fresh fruits and vegetables. Taylor & Francis Group, LLC, New York. doi:10.1201/9781420038934.ch2

Martin ST, Beelman RB (1996) Growth and enterotoxin production of *Staphylococcus aureus* in fresk packed mushrooms (*Agarcusbisporus*). J Food Prot 59(8):819–826

Moon JS, Lee AR, Jaw SH, Kang HM, Joo YS, Park YH, Kim MN, Koo HC (2007) Comparison of antibiogram, staphylococcal enterotoxin productivity, and coagulase genotypes among *Staphylococcus aureus* isolated from animal and vegetable source in Korea. J Food Prot 70 (11):2541–2548

Nguyen-The C, Carlin F (2000) Fresh and processed vegetables. In: Lund B, Baird-Parker T, Gould GW (eds) The microbial safety and quality of foods. Aspen Publishers Inc., Gaithesburg

Nipa MN, Mazumdar RM, Hasan MM, Fakruddin MD, Islam S, Bhuyan HR, Iqbal A (2011) Prevalence of multi drug resistant bacteria on raw salad vegetables sold in major markets of Chitagong City, Bangladesh. Middle-East J Sci Res 10(1):70–77

Normanno G, Corrente M, La Salandra G, Dambrosio A, Quagilia NC, Parisi A, Greco G, Bellacicco, AL, Virgilio S, Celano GV (2007) Methicillin-resistant *Staphylococcus aureus* (MRSA) in food animal origin produced in Italy. Int J Food Microbiol 117(2):219–222. doi:10. 1016/j.ijfoodmicro.2007.04.006

O'Brien S, Mitchel RT, Gillespie J, Adak GK (2000) The microbiological status of ready eat fruit and vegetables. Discussion Paper ACM/476, the Advisory Committee on the Microbiological Safety of Food (ACMSF), Food Standards Agency, The Stationary Office, London

Ofor MO, Okorie VC, Ibeawuchi II, Ihejirika GO, Obilo OP, Dialoke SA (2009) Microbial contaminants in fresh tomato wash water and food safety considerations in South-Eastern Nigeria. Life Sci J 6(3):80–82. Available http://www.sciencepub.net/life/0603/12_1197_microbial_life0603_80-82.pdf. Accessed 18 Oct 2016

O'Neill E, Pozzi C, Houston P, Smyth D, Humphreys H, Robinson DA, O'Gara JP (2007) Association between methicillin susceptibility and biofilm regulation in *Staphylococcus aureus* isolates from device related infection. J ClinBacteriol 45(5):1379–1388. doi:10.1128/JCM. 02280-06

Rich M, Roberts L (2004) Methycillin-resistant *Staphylococcus aureus* isolates from companion animals. Vet Rec 154(10):310

Rohde H, Frankenberger S, Zahringer U, Mack D (2010) Structure. Function and contribution of polysaccharide intercellular adhesion (PIA) to *Staphylococcus epidermidis* biofilm formation and pathogenesis of biomaterial-associated infection. Eur J Cell Biol 89(1):103–111. doi:10. 1016/j.ejcb.2009.10.005

Silbernagel KM, Jechorek RP, Carver CN, Horter BL, Lindberg KG (2003) 3M petrifilm staph express count plate method for the enumeration of *Staphylococcus aureus* in selected types of processed and prepared foods: collaborative study. J AOAC Int 86(5):954–962

Splittsoesser DF, Queale DT, Bowers JL, Wilkinson M (1980) Coliform content of frozen blanched vegetables packed in the United States. J Food Saf 2(1):1–11. doi:10.1111/j.1745-4565.1980.tb00386.x

Steinka I (2012) Jakosc mikrobiologiczna wybranych mrozonek warzywnych. Zeszyty Naukowe, Akademii Morskiej w Gdyni 72:55–59

Steinka I, Janczy A (2013) Ocena antybiotykooopornosci *Staphylococcus aureus* izolowanych z miesa mielonego. Bromat Chem Toksykol 66(2):211–215

Steinka I, Janczy A (2014) Badanie obecnosci antybiotykoopornych grzybow drozdzopodobnych w miesie mielonym. Przegl Hig Epidemiol 95(1):192–195. Available http://phie.pl/pdf/phe-2014/phe-2014-1-192.pdf. Accessed 18 Oct 2016

Tournas VH (2005a) Mould and yeast in fresh and minimally processed vegetables and sprouts. J Food Microbiol 99(1):71–77. doi:10.1016/j.ijfoodmicro.2004.08.009

TournasVH (2005b) Spoilage of vegetable crops by bacteria and fungi and related health hazards. Crit Rev Microbiol 31(1):33–44. doi:10.1080/10408410590886024

Viswanathan P, Kaur R (2001) Prevalence and growth of pathogens on salad vegetables, fruits and sprouts. Int J Hig Environ Health 203(3):205–213. doi:10.1078/S1438-4639(04)70030-9

Zhuang H, Barth MM, Hankison TR (2003) Microbial safety, quality and sensory aspects of fresh —cut fruits and vegetables: In: Novak JS, Sapers GM, Juneja VK (eds) Microbial safety of minimal processed foods. CRC Press, Boca Raton

References

Chapter 2
Technology and Chemical Features of Frozen Vegetables

Izabela Steinka, Caterina Barone, Salvatore Parisi and Marina Micali

Abstract The aim of this study has been the description of the current state of the art of frozen vegetables. One of the most promising food and beverage categories in the current market is represented by frozen products, although the modern food industry is officially born in 1928 in the United States of America. Before this date, previous freezing systems were based on the production of ice, with the use of refrigerating machines and the development of storage rooms. At present, the evolution of this sector can be briefly identified with the improvement of freezing techniques, the notable demand of food supplies worldwide, and the increasing number of frozen food typologies, including fruits and vegetables. Basically, frozen foods are very similar to original products when speaking of sensorial features. On the other side, some defects have been observed and correlated with freezing techniques and blanching treatments. The most used systems—air blast, plate and immersion freezers—are discussed with the description of correlated advantages and risks, including economic evaluations.

Keywords Air blast freezer · Cryogenic freezing · Frozen vegetable · Immersion freezer · Individual quick freezing · Plate freezer · Texture

Abbreviations

IQF   Individual quick freezing
INRAN   Istituto Nazionale di Ricerca per gli Alimenti e la Nutrizione

2.1 Introduction

One of the most promising food and beverage categories in the entire food market worldwide is represented by frozen products (Mallett 1993). Actually, it should be highlighted that the modern food industry is officially born in 1928 in the United States of America (Persson and Londahl 1993) with the introduction of double-belt contact freezers. Before this date, previous freezing systems were based on the

© The Author(s) 2017   23
I. Steinka et al., *The Chemistry of Frozen Vegetables*, Chemistry of Foods,
DOI 10.1007/978-3-319-53932-4_2

production of ice, the use of refrigerating machines and the development of storage rooms (slow freezing techniques). Because of technological limitations related to the first technological procedures, only three main food categories—meat, butter and fish—were frozen. The real difference between old freezing systems and the current situation may be easily identified (Arthey 1993; Cortellino 2016; Fellows 2009; Hui 2006; James and James 2016; North and Lovatt 2006; Parreño and Torres 2016) as follows:

(a) The improvement of freezing techniques, with the introduction of multi-plate freezers, individual quick freezing systems and cryogenic methods
(b) The remarkable demand of food supplies worldwide without seasonal varia-tions, differently from the past
(c) The increased request of readily available foods in different regions of the industrialised world
(d) The increasing number of food typologies which can be treated with freezing techniques.

With relation to the last point, is should be considered that fruits and vegetables are one of the most important categories of frozen foods at present. Probably, this reflection may reveal the real difference between past and present times, when speaking of frozen technologies. Moreover, fruits and vegetables are considered with great favour when speaking of good nutritional advices and the necessity of supplying a good amount of vitamins, minerals, fibres, and other healthy principles with anti-oxidant properties (Delgado et al. 2017). As an example, Italian Guidelines for a Safe Nutrition recommend five portion of vegetable products per day: in other terms, 600–800 g of vegetables should be consumed daily (INRAN 2003). On the other side, certain food processing technologies may cause peculiar damages to vegetable tissues and other foods (Parisi 2002). For all these reasons, the necessity of long-durability foods such as vegetables and fruits is 'mandatory' in globalised markets (Parisi 2016). Finally, the perishability of food commodities —all food commodities—should not be forgotten. Parisi has enunciated recently his Law of Food Degradation: in brief, there are not foods or beverage which could remain unmodified during time; chemical, physical, microbiological and structural properties are always subjected to modifications, without exceptions (Parisi 2002; Volpe et al. 2015).

2.2   Frozen Foods: Chemical and Physical Modifications

Basically, frozen foods are similar to original products when speaking of sensorial features. In some situation, frozen foods may exhibit similar or ameliorated organoleptic features (Lisiewska and Kmiecik 2000), although other authors signal different properties (Le Bail et al. 2016). Anyway, basic sensorial properties are judged satisfactory in the most part of situations; sometimes, these results are correlated with good or excellent raw materials (Senesi 1984), the use of particular additives such as maltodextrins (Specter and Setser 1994) and the reduction of lipidic oxidation in peculiar foods (Erickson 1997). On the other side, the modification of chemical parameters for frozen foods during storage could give some unexpected surprise when speaking of fish products (Arannilewa et al. 2006).

In general, chemical features and sensorial properties of frozen foods can suffer the following modifications:

1. The formation of ice crystals into foods under freezing processes is well known and extensively studied. The problem correlated with ice crystals is not dependent on the crystallisation of aqueous molecules, but with the destruction of tissues because of large ice crystals and the possible emersion of ice needles from food surfaces, similar to the situation observed in deep-frozen cheeses (Parisi et al. 2016). This phenomenon should be carefully discussed in detail in terms of kinetics (Belitz et al. 2009; Bevilacqua and Zaritzky 1982; Donhowe et al. 1991; Lévy et al. 1999). By the microscopic viewpoint, it could be affirmed that the obvious removal of liquid water molecules is cause of partial protein dehydration with consequent spatial modifications of lipoproteins. With particular reference to meat products, the formation of large ice crystals (and consequent macroscopic damages) is inhibited on condition that freezing is very rapid. With relation to physical modifications, a certain variation of rheological properties in intermediate liquid masses can be observed (Belitz et al. 2009). The problem is correlated with the amount of non-freezing water even at −30 °C, the high viscosity of the remaining liquid medium, and the consequent slowing of freezing (Walstra 2003)

2. Should large ice crystals be formed because of too slow freezing in meats, myofibrillar proteins could be irreversibly modified because of the increasing salt amount (Belitz et al. 2009). Anyway, volumetric increase has to be expected (Walstra 2003)

3. Certain food products may be concentrated with freezing techniques on condition that the initial moisture amount is remarkable. The resulting product is a network of glassy compounds with dispersed ice crystals. The process can give very stable foods on condition that the 'special glass transition temperature' is kept continually (Walstra 2003). This thermal value is generally between −10 and −40 °C: should freezing be carried in this thermal range, deep-frozen foods would not show physical or chemical variations with some negligible exception (Walstra 2003)

4. The diminution in solubility of fish proteins has been reported during frozen storage in certain situations involving Maillard reactions. In addition, deep-freezing techniques inhibit proteolysis in fish and may cause some colorimetric variation and textural changes in certain products such as whale meat (Belitz et al. 2009). Consequently, should proteolytic reactions be observed in frozen fish, the reason would be easily found in the quality of original fish

5. Some decrease in vitamin amounts has been observed in cooked and drained frozen products

6. The possible loss of aroma in certain frozen meat and fish products has been observed and 'amended' by means of the addition of monosodium glutamate. This correction can be also used when speaking of processed and canned fish and meat products.

These chemical and physical modifications concern all possible food typologies in general. With concern to vegetable products, it has been reported that blanched vegetables may exhibit a certain loss of chlorophylls because of their transformation in pheophytins even at -18 °C (Belitz et al. 2009). The same thing can be told when speaking of certain vitamins such as vitamin C. Reasons for lowered amounts are not linked with incorrect freezing methods, but with blanching treatment and storage. However, the most important problem for these products concerns always irreversible textural modifications (Belitz et al. 2009) and general damages to plant tissues after defrosting (Aked 2000): decrease in vitamins; flavour modifications (caused by residual enzymatic activity); discoloration; and microbial spreading by yeast (if thermal values are different from expected values). The next section discusses main technological solutions for vegetable and other food products.

2.3   Frozen Foods: Main Technological Processes

In general, freezing techniques can be carried out with the use of two different freezers types (Senesi 1984):

(a) Mechanical freezers
(b) Cryogenic freezers.

Actually, the modern industry is accustomed to use three main freezers typologies (Senesi 1984):

1. Air blast freezers
2. Plate freezers (also named contact freezers)
3. Immersion freezers (also named spray freezers).

A brief description of these systems is now provided in the following sections.

2.3.1   Air Blast Freezers

The most used freezer system appears to be the 'air blast freezer' type (Johnston et al. 1994). Briefly, this system is based on the forced convection of cold air into rooms of various dimensions by means of fans. In this way, regularly or irregularly shaped foods can be frozen with good results. Moreover, there are not peculiar failures linked to 'critical' dimensions or shapes, although a pre-determined air blast freezer is not easily modifiable after installation and initial trials. In other words, different shapes could require different freezers, and the preliminary design of air blast freezers is surely critical. Normally, air speed values could be 5 m/s although larger industries may prefer higher values (10–15 m/s) with the aim of reducing freezing times in continuous processes (Johnston et al. 1994).

On the other side, main risks depend mainly on the presence of physical obstacles. Packaging materials and other non-food materials have to be removed. For this reason, Individual Quick Freezing (IQF) systems are available at present. These machines can easily freeze unpackaged foods, and the frozen product can be immediately packaged just before IQF. Actually, the IQF system may show a notable problem when speaking of economic results: the possible superficial dehydration of products. For this reason, a correct design is critical; economic evaluations should be taken into account (Senesi 1984).

Anyway, air blast freezers can be defined as discontinuous- or continuous-flow machines, depending on excessive 'dead times' in production. Batch-continuous freezers are widely used nowadays; products are uploaded on proper belts (Johnston et al. 1994; Senesi 1984).

2.3.2   Plate Freezers

Differently from air blast freezers, contact freezers are not 'customisable' systems. Only regularly shaped foods can be treated with these machines (Johnston et al. 1994).

The real difference between available types is substantially the vertical or horizontal arrangement of freezing plates into the freezer. For this reason, available machines are named horizontal or vertical plate freezers. Basically, the heat removal is realised by means of the contact between food products and cold plates: a refrigerant fluid can flow into plates and maintain constant cold conditions. In addition, plates may be moveable during the process (Johnston et al. 1994). Unfortunately, the presence of packaging materials and excessive air amount into 'open spaces' between food products may delay freezing times.

2.3.3   Immersion Freezers

Differently from above-mentioned systems, immersion freezers are based on the direct contact of food products with a refrigerant fluid: nitrogen gas (temperatures: −50 to −196 °C) or liquefied carbon dioxide (temperature range: −50 to −70 °C). Apparently, this approach is excellent and should be recommended because of the limited size of freezers and the rapidity of freezing processes. On the other hand,

1. Supplies of refrigerant fluids such as required nitrogen or carbon dioxide may be a real problem in certain Countries

2. Carbon dioxide can be really dangerous; consequently, should a similar system use this material, adequate safety countermeasures (example: ventilation) would be mandatory
3. Frozen products could be easily damaged because of high freezing speed
4. Anyway, economic costs are convenient on condition that discontinuous productions are considered.

References

Aked J (2000) Fruits and vegetables. In: Kilcast D, Subramaniam P (eds) The stability and shelf-life of food. Woodhead Publishing Limited, Cambridge

Arannilewa ST, Salawu SO, Sorungbe AA, Ola-Salawu BB (2006) Effect of frozen period on the chemical, microbiological and sensory quality of frozen tilapia fish (Sarotherodungaliaenus). Nutr Health 18(2):185–192. doi: 10.1177/026010600601800210

Arthey D (1993) Freezing of vegetables and fruits. In: Mallett CP (ed) Frozen food technology. Blackie Academic & Professional, Glasgow

Belitz HD, Grosch W, Schieberle P (2009) Food chemistry, 4th edn. Springer, Berlin

Bevilacqua AE, Zaritzky NE (1982) Ice recrystallization in frozen beef. J Food Sci 47(5):1410–1414. doi:10.1111/j.1365-2621.1982.tb04950.x

Cortellino G (2016) Quality and safety of frozen fruits. In: Sun DW (ed) Handbook of frozen food processing and packaging, 2nd edn. CRC Press, Boca Raton

Delgado AM, Almeida MDV, Parisi S (2017) Chemistry of the Mediterranean diet. Springer International Publishing, Cham

Donhowe DP, Hartel RW, Bradley RL (1991) Determination of ice crystal size distributions in frozen desserts. J Dairy Sci 74(10):3334–3344. doi:10.3168/jds.S0022-0302(91)78521-4

Erickson MC (1997) Lipid oxidation: flavor and nutritional quality deterioration in frozen foods. In: Erickson MC, Hung YC(1997) Quality in frozen foods. Springer, New York. doi:10.1007/978-1-4615-5975-7_9

Fellows PJ (2009) Food processing technology: principles and practice, 3rd edn. Woodhead Publishing, Oxford, Cambridge and New Delhi

Hui YH (2006) Handbook of food science, technology, and engineering, vol 4. CRC Press, Boca Raton

INRAN (2003) Linee Guida per una Sana Alimentazione Italiana, Revisione 2003. Ministero Politiche Agricole e Forestali, Rome, and Istituto Nazionale di Ricerca per gli Alimenti e la Nutrizione (INRAN), Rome. Available http://www.salute.gov.it/portale/documentazione/p6_2_2_1.jsp?id=652. Accessed 09 Nov 2016

James SS, James C (2016) Quality and safety of frozen meat and meat products. In: Sun DW (ed) Handbook of frozen food processing and packaging, 2nd edn. CRC Press, Boca Raton

Johnston WA, Nicholson FJ, Roger A, Stroud GD (1994) Freezing and refrigerated storage in fisheries. FAO Fisheries technical paper—40. Food and Agriculture Organization of the United Nations, Rome. Available http://www.fao.org/docrep/003/v3630e/V3630E00.htm#Contents. Accessed 09 Nov 2016

Le Bail A, Tzia C, Giannou V (2016) Quality and safety of frozen bakery products. In: Sun DW (ed) Handbook of frozen food processing and packaging, 2nd edn. CRC Press, Boca Raton

Lévy J, Dumay E, Kolodziejczyk E, Cheftel JC (1999) Freezing kinetics of a model oil-in-water emulsion under high pressure or by pressure release. Impact on ice crystals and oil droplets. LWT-Food Sci Technol 32(7):396–405. doi:10.1006/fstl.1999.0567

Lisiewska Z, Kmiecik W (2000) Effect of storage period and temperature on the chemical composition and organoleptic quality of frozen tomato cubes. Food Chem 70(2):167–173. doi:10.1016/S0956-7135(99)00110-3

Mallett CP (1993) Editorial introduction. In: Mallett CP (ed) Frozen food technology. Blackie Academic & Professional, Glasgow

North MF, Lovatt SJ (2006) Freezing methods and equipment. In: Sun DW (ed) Handbook of frozen food processing and packaging. CRC Press, Boca Raton

Parisi S (2002) I fondamenti del calcolo della data di scadenza degli alimenti: principi ed applicazioni. Ind Aliment 41(417):905–919

Parisi S (2016) The world of foods and beverages today. Online video course, Learning.ly/The Economist Group, 750 Third Ave NY, NY 10017. Available http://learning.ly/products/the-world-of-foods-and-beverages-today-globalization-crisis-management-and-future-perspectives Accessed 09 Nov 2016

Parisi S, Delia S, Laganà P (2004) Il calcolo della data di scadenza degli alimenti: la funzione Shelf Life e la propagazione degli errori sperimentali. Ind Aliment 43:735–749

Parisi S, Barone C, Caruso G (2016) Packaging failures in frozen curds. The use of a BASIC software for mobile devices. Food Packag Bull 25(1 & 2):16–19

Parreño WC, Torres MDÁ (2016) Quality and safety of frozen vegetables. In: Sun DW (ed) Handbook of frozen food processing and packaging, 2nd edn. CRC Press, Boca Raton

Persson PO, Londahl G (1993) Freezing technology In: Mallett CP (ed) Frozen food technology. Blackie Academic & Professional, Glasgow

Senesi E (1984) Surgelazione dei prodotti vegetali. In: Ottaviani F (ed) Microbiologia dei prodotti di origine vegetale – ecologia ed analisi microbiologica. Chiriotti Editori, Pinerolo

Specter SE, Setser CS (1994) Sensory and physical properties of a reduced-calorie frozen dessert system made with milk fat and sucrose substitutes. J Dairy Sci 77(3):708–717. doi:10.3168/jds. S0022-0302(94)77004-1

Volpe MG, Di Stasio M, Paolucci M (2015) Polymers for food shelf-life extension. In: Cirillo G, SpizzirriUG, Iemma F (eds) Functional polymers in food science: from technology to biology, vol 1: food packaging. Scrivener Publishing, Beverly

Walstra P (2003) Physical chemistry of foods. Marcel Dekker Inc., New York

Chapter 3
Instrumental Systems for the Control of Frozen Vegetables During Refrigeration

Izabela Steinka, Caterina Barone, Salvatore Parisi and Marina Micali

Abstract The aim of this study has been the description of storage technologies used at present for frozen foods. The necessity of long-durability foods seems one of the main distinctive features of the globalised market. In addition, the perishability of certain food commodities has to be considered as a first-level pillar of the modern food science and technology. There are not foods or beverages which could remain unmodified during time, on the basis of Parisi's Law of Food Degradation. A reliable freezing process is only the first step in the entire preservation flow chart with the exclusion of initial production steps; in fact, the most important result is the final performance of deep-frozen products until the desired or planned expiration date. For this reason, frozen foods have to remain continually exposed to low temperatures (≤ -18 °C). This condition is the so-called 'cooling chain' and concerns logistic operators into and outside food companies until the final use. The current availability of system controls after freezing and packaging steps—freezer warehouses, rooms, chillers and household freezers—is notable enough in the current market of industrial equipments. Each of these systems is described here with correlated advantages and cost evaluations.

Keywords Chest freezer · Freezer warehouse · Frozen storage · Frozen vegetable · Household freezer · Logistics · Open-top freezer · Volumetric dimension

3.1 The Importance of Frozen Storage: Introduction

The necessity of long-durability foods seems one of the main distinctive features of the globalised market (Parisi 2016). In addition, the perishability of certain food commodities has to be considered as a first-level pillar of the modern food science and technology. Parisi had enunciated recently his Law of Food Degradation (Parisi 2002a; Volpe et al. 2015): in brief, there are no foods or beverage which could remain unmodified during time; chemical, physical, microbiological and structural properties are always subjected to modifications, without exceptions.

© The Author(s) 2017

I. Steinka et al., *The Chemistry of Frozen Vegetables*, Chemistry of Foods,
DOI 10.1007/978-3-319-53932-4_3

With specific relation to frozen foods, the performance of freezing techniques cannot be assured without reliable instruments for the control of vegetables during the critical 'storage' step. Basically, freezing technologies can be helpful if the main goal of preservation is the more or less rapid freezing of food products, including advantages and risks. Generally, each technological system—air blast freezers, plate freezers, cryogenic freezers—has some weak point when speaking of dimensional (volumetric) capability, possible airflow discontinuities into operating rooms, and excessive costs (Sect. 2.3). However, a reliable freezing process is only the first step in the entire preservation flow chart with the exclusion of initial production steps (choice of raw materials, washing and preliminary heat treatments). In fact, the most important result is the final performance of deep-frozen products, including vegetables, until the desired or planned expiration date. For this reason, frozen foods have to remain continually exposed to low temperatures (≤ -18 °C). This condition is the so-called 'cooling chain' and concerns logistic operators into and outside food companies until the final consumption.

Basically, the flow chart of frozen products could be also defined as the interaction between different key players (Fig. 3.1). In this way, it can be easily seen that (Vanhaverbeke et al. 2008):

Frozen foods and Key Players in the Food Chain

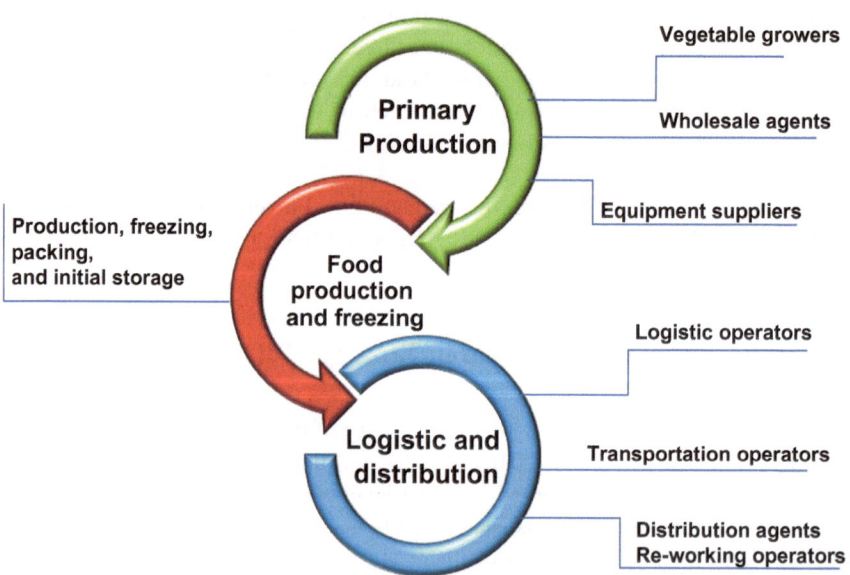

Fig. 3.1 A general flow chart for frozen vegetable products, based on the description of key players

1. The first group of food players—vegetable growers, wholesale agents and equipment suppliers—does not carry out freezing processes, although some exception is possible
2. Producers of frozen vegetables perform the following steps: production, freezing, packing and initial storage of products. These actors can also sale their own products. Each of these steps can be carried out by a different company, although technological solutions seem to favour the management of production, freezing, packaging and initial storage steps in the same place
3. The third group of external agents perform logistic operations, including also transport and continuous storage at temperatures ≤ -18 °C, and the final distribution and/or re-working of selected foods and beverages. Consumers are not part of this category, while customers can be catering companies, retail services and other food companies. These actors may correspond to the same commercial player.

Substantially, the third group of food players is involved in different activities taking a very long temporal period, when speaking of frozen foods. In detail, expiration dates are fixed generally several months after the final manufacturing date (freezing and packaging steps), but the estimation can depend on many factors (Fu and Labuza 1997), including the food typology. In fact, different frozen foods can show different results with dissimilar failures, and frozen vegetables are surely one of the most interesting products in this ambit. Several data concerning frozen vegetables have shown that shelf-life intervals may be comprised between 175 and 720 days for certain frozen vegetables, depending on the peculiar product (Guadagni 1968). By the commercial viewpoint, storage periods for many frozen fruits and vegetables may reach 12 months, although storage temperatures such −30 °C may guarantee more months. Anyway, the reliability of storage controls with relation to thermal values remains critical.

With reference to frozen vegetables, the availability of system controls after freezing and packaging steps is notable enough in the current market of industrial equipments.

3.2 Industrial Freezers for Frozen Storage

It should be considered that the main risk for frozen vegetables (and normal foods in the same storage condition) is the possible occurrence of microbial contamination by pathogenic agents (Parisi 2002b). On these bases, freezing techniques can assure a good level of food hygiene and safety. However, vegetative microorganisms may start again with their normal activities if thermal values increase, and this risk can surely become a concern if higher thermal values are observed for extended temporal periods (Parisi 2002a). For this reason, the cooling chain has to remain constantly under −10 °C at least. Some problem can be correlated also to poor packaging performances (Parisi 2012, 2013), but microbiological concerns are mainly dependent on thermal abuses.

Basically, the 'thermal history' of frozen products during storage may be influenced by variations of recorded thermal values in the storage period. In particular, thermal variations and the possibility of thermal gradients into a single food unit may be cause of moisture loss: in other words, water molecules may migrate from one section of the food product to another section in function of thermal values with undesired displacement of the aqueous mass and possible re-crystallisation. These risks may be observed if air thermal fluctuations are >2 °C. As a result, textural and other sensorial features may be modified (Barbosa-Cánovas et al. 2005; Blanshard and Franks 1987).

For all these reasons, cold storage should be maintained constant with very low thermal values. Ideally, storage temperatures should reach −30 °C (Johnston et al. 1994), but similar performances are often correlated with the volumetric dimension of storage rooms and the maximum amount of storable foods. In addition, expensiveness has to be considered when speaking of the simple maintenance and needed electric power for maintaining constantly low temperatures (Barbosa-Cánovas et al. 2005). In fact, freezing and storage steps may represent about 10% of the total production and storage costs (Persson and Londahl 1993). Consequently, freezing is important but storage may have an equal importance. In general, the basic requirement is that frozen foods have to be stored at −18 °C or lower temperatures.

Another important pre-requisite for storage conditions of frozen foods and vegetables in particular is the monitoring of relative humidity. The reason is correlated with possible food drying episodes, when packaging materials do not assure good protection (Parisi 2013).

Because of the above-discussed critical points in the management of frozen foods (in a general way), a simplified description of instruments for thermal controls in the cooling chain (logistic operators, transportation vehicles, sea containers, mass retailers) is required (Micali et al. 2009). These systems can be described as follows:

1. One-level and multi-level freezer warehouses; minimum volumetric dimensions may be 500 m^3 when speaking of fish products (Johnston et al. 1994). In general, lower volumetric capacity is convenient; however, the current trend is in favour of larger warehouses. Ammonia is the preferred refrigerant fluid for these systems

2. Freezer rooms. Differently from freezer warehouses, these systems are destined to lower amounts of frozen foods. In some situations, warehouses are subdivided in different chillers with the aim of storing different foods in different and isolated areas, for safety and logistic purposes. Ammonia is the preferred refrigerant fluid for these systems. A category apart is represented by freezer units for transportation purposes on trucks, sea containers and flights. These equipments may be very different and are generally required to lower inner temperatures in a little temporal period (example: freezers for sea containers should arrive to desired temperatures within 24–36 h). The interested Reader is

invited to consult more specific literature concerning the description of freezers
for transportation purposes

3. Open-top and chest chillers. These average—and low—volume freezers are
   specifically used for food exposition near mass retailers and household storage
   purposes, respectively. With reference to distribution services, open-top freezers
   can be easily opened and give fully accessibility to frozen products by means of
   a simple transparent cover. Other solutions, including chest freezers, are represented by glass-lidded systems because of a simple sliding glass lid.
   Household freezers are similar. In general, basic requirements for these average
   and little-sized machines are reliable control of inner temperature, with continuous record and thermal display; good inner visibility; easy accessibility to
   frozen products (logistic workers are required to take their operations easily);
   low energy costs; and the use of low-heating fluorescent lamps for lighting
   purposes (heat sources have to be completely removed)

4. Household freezers/chillers. Normal systems, widely diffused, have a com-
   pressor unit only and two different sections at +4 and −18 °C, respectively. The
   second of these sections, reserved for frozen products, does not exceed one
   quarter of the entire volumetric capacity. The current trend is in favour of
   chillers with two refrigerating units and two sections (+4 and −18 °C) with
   similar volumetric dimensions.

References

Barbosa-Cánovas GV, Altunakar B, Mejía-Lorío DJ (2005) Freezing of fruits and vegetables. An
    agribusiness alternative for rural and semi-rural areas. FAO Agricultural Services Bulletin—
    158. Food and Agriculture Organization of United Nations (FAO), Rome

Blanshard JMV, Franks F (1987) Ice crystallization and its control in frozen food systems. In:
    Blanshard JMV, Lillford P (eds) Food structure and behaviour. Academic Press, London,
    pp 51–65

Fu B, Labuza TP (1997) Shelf life testing: procedures and prediction methods for frozen foods. In:
    Erickson MC, Hung YC(1997) Quality in frozen foods. Springer, New York. doi:10.1007/978-
    1-4615-5975-7_19

Guadagni DG (1968) Cold storage life of frozen foods and vegetable as a function of time and
    temperature. In: Hawthorne J, Rolfe E (eds) Low temperature biology of food stuffs. Pergamon
    Press, Oxford, pp 399–412

Johnston WA, Nicholson FJ, Roger A, Stroud GD (1994) Freezing and refrigerated storage in
    fisheries. FAO Fisheries technical paper—340. Food and Agriculture Organization of the
    United Nations, Rome. Available http://www.fao.org/docrep/003/v3630e/V3630E00.
    htm#Contents. Accessed 09 Nov 2016

Micali M, Parisi S, Minutoli E, Delia S, Laganà P (2009) Alimenti confezionati e atmosfera modificata.
    Caratteristiche basilari, nuove procedure, applicazioni pratiche. Ind Aliment 48:35–43

Parisi S (2002a) I fondamenti del calcolo della data di scadenza degli alimenti: principi ed
    applicazioni. Ind Aliment 41(417):905–919

Parisi S (2002b) Profili evolutivi dei contenuti batterici e chimico-fisici in prodotti lattiero-caseari.
    Ind Aliment 41(412):295–306

Parisi S (2012) Food packaging and food alterations: the user-oriented approach. Smithers Rapra Technology, Shawbury

Parisi S (2013) Food industry and packaging materials—performance-oriented guidelines for users. Smithers Rapra Technology, Shawbury

Parisi S (2016) The world of foods and beverages today. Online video course, Learning.ly/The Economist Group, 750 Third Ave NY, NY 10017. Available http://learning.ly/products/the-world-of-foods-and-beverages-today-globalization-crisis-management-and-future-perspectives. Accessed 09 Nov 2016

Persson PO, Londahl G (1993) Freezing technology In: Mallett CP (ed) Frozen food technology. Blackie Academic & Professional, Glasgow

Vanhaverbeke W, Larosse J, Winnen W (2008) The Flemish frozen-vegetable industry as an example of cluster analysis. In: Hulsink W, Dons H (eds) Pathways to high-tech valleys and research triangles: innovative entrepreneurship, knowledge transfer and cluster formation in Europe and the United States. Springer, Netherlands, pp 249–274

Volpe MG, Di Stasio M, Paolucci M (2015) Polymers for food shelf-life extension. In: Cirillo G, Spizzirri UG, Iemma F (eds) Functional polymers in food science: from technology to biology, vol 1: food packaging. Scrivener Publishing, Beverly

Chapter 4
Colorimetric Modifications in Frozen Vegetables

Izabela Steinka, Caterina Barone, Salvatore Parisi and Marina Micali

Abstract The success of frozen foods in the current market is linked to the virtual absence of sensorial modifications, although irreversible textural modifications and general damages to plant tissues after defrosting are reported when speaking of vegetables. In addition, flavour changes and discoloration have been reported. Reasons may be different and possibly linked to blanching treatment and storage, while freezing techniques could be not so important. With exclusive relation to colorimetric changes, the main cause is reported to be the residual activity of enzymes such as lipoxygenase and peroxidase. For these reasons, the limitation of available oxygen during freezing is a good strategy. The so-called 'freezer burn' is also considered as a probable cause. In addition, physical browning may occur after mechanical manipulation and correlated stress, as stated by Parisi in his second Law of Food Degradation. The analytical evaluation of colour modification may be carried out by means of CIE lab-based colorimetric systems, or 'digital image analysis and processing' techniques. Moreover, enzymatic browning may be evaluated spectrophotometrically. The diminution of light-sensible vitamin C can be also considered. Finally, food industries are accustomed to use simplified food pictures or colorimetric charts with the aim of simplifying inner quality controls.

Keywords Blanching · Discoloration · Enzymatic browning · Flavour · Freezer burn · Frozen vegetable · Lipoxygenase · Peroxidase · Texture · Vitamin C

Abbreviations

CIE   Commission Internationale de l'Éclairage
DIAP   Digital image analysis and processing

© The Author(s) 2017   37
I. Steinka et al., *The Chemistry of Frozen Vegetables*, Chemistry of Foods,
DOI 10.1007/978-3-319-53932-4_4

4.1 Colour Modifications in Frozen Vegetables: An Introduction

The success of frozen foods in the current market is necessarily linked to the substantial absence of serious modifications when speaking of sensorial features. Actually, this affirmation may be questionable in certain situations when speaking of vegetables (Sect. 2.2). In fact, the most important problem for frozen vegetable concerns always irreversible textural modifications and general damages to plant tissues after defrosting. In general, several possible organoleptic modifications may be observed in frozen vegetables, including flavour changes and discoloration. Reasons may be different and possibly linked to blanching treatment and storage, while freezing techniques could be not so important when speaking of process failures.

Basically, sensorial properties of vegetables such as flavour and colour are negatively influenced by enzymatic activities. Because of the critical importance of quality of vegetable raw materials (microbial ecology, freshness, etc.), freezing processed could only limit the speed of possible advanced enzymatic activities. In other words, freezing methods can surely delay the negative result of enzymatic reactions and consequent food degradation, but the complete eradication of this important problem is impossible (Martinez et al. 2004). For these reasons, blanching treatments are needed with the aim of preventing biochemical reactions (Sect. 2.3).

Anyway, it has been reported that frozen vegetables retain basic sensorial features such as flavours and colours with good performances if compared with differently processed vegetables (Martinez et al. 2004).

By the viewpoint of normal consumers, flavour features of frozen vegetables could be appreciated after defrosting and preparation, while colorimetric appearance of the same products could be judged before defrosting, although this behaviour is not recommended. However, a certain correlation between colours of frozen products before defrosting and the final result might be expected. As a result, consumers may consider colorimetric appearance of frozen vegetables as a distinctive key point for these products, in terms of purchase evaluation. The same thing should be expected in general for all food products (Martín-Diana et al. 2007).

For these reasons, a brief discussion concerning the importance of discolouration phenomena in frozen foods should be shown in this book.

4.2 Colour Modifications in Frozen Vegetables. Main Reasons

By a general viewpoint, it can be affirmed that the main causes of discolouration in frozen vegetables are ascribed to the residual activity of enzymes such as lipoxygenase, peroxidase and other well-known active biomolecules such as catalase and polyphenoloxidase (Martinez et al. 2004). In particular, lipoxygenase and peroxidase have been recognised probably responsible for many discolouration phenomena in frozen vegetables during storage at −18 °C and lower temperatures (Martinez et al. 2004). Consequently, it has been suggested that blanching treatments should delay the remaining enzymatic activity in these foods (Robinson 1991). On the other side, colorimetric variations may be negligible in certain vegetables under frozen storage until one year after freezing (Brewer et al. 1994). For these reasons, the analytical estimation of residual peroxidase and lipoxygenase activity in frozen vegetables has been recommended with a desired 95% level of inactivation for the first enzyme (Martinez et al. 2004).

Chemically, discolouration is the modification of natural pigments belonging to chlorophylls and carotenoids, although enzymatic reactions can also cause evident browning defects. Exposure to light can surely worsen the situation (Kilcast and Subramaniam 2000), while thermal influence is supposed to be negligible under correct frozen storage. With relation to green-coloured vegetables, the greenish appearance is the result of prevailing chlorophyll a and b molecules in comparison with pheophytins a and b. Substantially, chlorophylls (green colour) can be degraded (Canjura et al. 1991) and turned into pheophytins (brown appearance). In addition, the concomitant action of residual lipoxydase may worsen the visual result in favour of brownish colours (Martinez et al. 2004).

With reference to red-coloured vegetables, the observed enzymatic browning may be caused by polyphenoloxidase with oxygen availability. In these conditions, quinines may be produced and subsequently attack ascorbic acid and anthocyanins, responsible for red colours. In-progress enzymatic reactions can turn original appearance into brownish colours with unacceptable results for consumers (Martinez et al. 2004).

For these reasons, a good preservation strategy can be the limitation of available oxygen into packaged foods by means of adequate wrapping materials or air removal before during freezing procedures, where possible (Sect. 3.2). The so-called 'freezer burn'—the localised appearance of brown or white spots on vegetable surfaces as the result of moisture loss during freezing—should also be considered.

Finally, it should be considered that colorimetric changes may be also caused by physical browning after mechanical manipulation and correlated stress (Martín-Diana et al. 2007), as stated by Parisi in his second Law of Food Degradation (Parisi 2002).

4.3 Colour Modifications in Frozen Vegetables. Analytical Evaluation

In general, the most used strategy for evaluating colorimetric appearance is based on colorimetric 'Commission Internationale de l'Éclairage' (CIE) Lab systems (CIE 1978; Martinez et al. 2004). In the CIE lab approach, a colour space in three dimensions is identified with three basic parameters: lightness or L*, and chromaticity coordinates a* and b*, where a* can show negative (green direction) or positive (red direction) values, and b* may express negative (blue colour) or positive results (yellow colour). A distinctive disadvantage of this technique is the lack of colour uniformity when speaking of many single vegetable products (Aked 2000).

It has been also reported that colorimetric evaluations may be carried out by means of the analytical determination of browning-related enzymes (spectropho- tometric analysis), including peroxidase and polyphenol oxidase. Interestingly, the importance of these enzymes appears correlated with their localisation in vascular or photosynthetic tissues; consequently, analyses could be recommended with relation to photosynthetic tissues only (Martín-Diana et al. 2007). In addition, it has been suggested that polyphenol oxidase synthesis can be reduced or inhibited by means of the use of powerful oxidants; actually, the main reason of observed results appears to be dependent on the reduction of microbial activity and the lower res- piration amount (Micali et al. 2009). The addition of food additives such as sodium acid pyrophosphate to blanching water is sometimes recommended with relation to discolouration phenomena.

Another simple strategy concerns the monitoring of light-sensible vitamin C in certain vegetables such as green peas (Giannakourou and Taoukis 2003). Similar to the determination of residual enzymes, the basic aim is the kinetic study of colour degradation in examined vegetables versus time under frozen storage. At the same time, experiments considering also CIE lab parameters could be very useful in certain situations because of the demonstrated correlation between chemical degradation and colour parameters, including L*.

Food industries are accustomed to use simplified food pictures or colorimetric charts with the aim of simplifying inner quality controls (Aked 2000; Kilcast 2000). This approach can be very useful in certain ambits and possibly correlated with analytical determination, provided that a certain number of different samples have been examined, similar to other studies for non-frozen products (Parisi et al. 2013). In fact, it has been reported that colour unacceptability may be assumed as a quality index for frozen vegetables (Giannakourou and Taoukis 2003).

Finally, the examination of frozen foods can be also carried out by means of 'digital image analysis and processing' (DIAP) techniques (Parisi 2013): these techniques concern the analysis of digital pictures with interesting applications, including also non-frozen products (Parisi et al. 2013). On the other hand, stan- dardised DIAP procedures are surely needed in this ambit (Kilcast 2000; Parisi 2013).

References

Aked J (2000) Fruits and vegetables. In: Kilcast D, Subramaniam P (eds) The stability and shelf-life of food. Woodhead Publishing Limited, Cambridge
Barbosa-Cánovas GV, Altunakar B, Mejía-Lorío DJ (2005) Freezing of fruits and vegetables. An agribusiness alternative for rural and semi-rural areas. FAO Agricultural Services Bulletin— 158. Food and Agriculture Organization of United Nations (FAO), Rome

Brewer MS, Klein BP, Rastogi BK, Perry AK (1994) Microwave blanching effects on chemical, sensory and color characteristics of frozen green beans. J Food Qual 17(3):245–259. doi:10. 1111/j.1745-4557.1994.tb00147.x

Canjura FL, Schwartz SJ, Nunes RV (1991) Degradation kinetics of chlorophylls and chlorophyllides. J Food Sci 56(6):1639–1643. doi:10.1111/j.1365-2621.1991.tb08660.x

CIE (1978) Recommendations on uniform color spaces, color-difference equations, psychometric color terms. Supplement N. 2 to CIE publication N. 15. Commission Internationale de l'Eclairage (CIE), Paris

Giannakourou MC, Taoukis PS (2003) Application of a TTI-based distribution management system for quality optimization of frozen vegetables at the consumer end. J Food Sci 68 (1):201–209. doi:10.1111/j.1365-2621.2003.tb14140.x

Gioffrè ME, Parisi S, Piccione D, Micali M, Delia S, Laganà P (2009) Raffronto tra analisi microbiologiche e valutazione organolettica degli alimenti. Casi di studio in diversi comparti alimentari. Ig San Pubbl n. 5/2009, Suppl:392

Kilcast D (2000) Sensory evaluation methods for shelf-life assessment. In: Kilcast D, Subramaniam P (eds) The stability and shelf-life of food. Woodhead Publishing Limited, Cambridge

Kilcast D, Subramaniam P (2000) Introduction. In: Kilcast D, Subramaniam P (eds) The stability and shelf-life of food. Woodhead Publishing Limited, Cambridge

Martín-Diana AB, Rico D, Frias JM, Barat JM, Henehan GTM, Barry-Ryan C (2007) Calcium for extending the shelf life of fresh whole and minimally processed fruits and vegetables: a review. Trends Food Sci Technol 18(4):210–218. doi:10.1016/j.tifs.2006.11.027

Martinez D, Romero SC, Valero D (2004) Quality control in frozen vegetables. In: Hui YH, Cornillon P, Guerrero Legaretta I, Lim MH, Murrell KD, Nip WK (eds) Handbook of frozen foods. Marcel Dekker, New York

Micali M, Parisi S, Minutoli E, Delia S, Laganà P (2009) Alimenti confezionati e atmosfera modificata. Caratteristiche basilari, nuove procedure, applicazioni pratiche. Ind Aliment 48: 35–43

Parisi S (2002) I fondamenti del calcolo della data di scadenza degli alimenti: principi ed applicazioni. Ind Aliment 41(417):905–919

Parisi S (2013) Food industry and packaging materials—performance-oriented guidelines for users. Smithers Rapra Technology, Shawbury

Parisi S, Laganà P, Gioffrè ME, Minutoli E, Delia S (2013) Problematiche Emergenti di Sicurezza Alimentare. Prodotti Etnici ed Autenticità. In: Proceedings of the XXIV Congresso Interregionale Siculo-Calabro SitI, Palermo, 21–23 June 2013. Euno Edizioni, Leonforte, pp 35–36

Robinson DS (1991) Peroxidases and their significance in fruits and vegetables. In: Fox BF (ed) Food enzymology, vol 1. Elsevier Applied Science, London, pp 399–426